CONTENTS

STARTING DESIGN & TECHNOLOGY

ENE ROL

CASSELL

Cassell Publishers Limited
Artillery House
Artillery Row
London SW1P 1RT

First published 1989

ISBN 0-304-31645-8

Typeset by Flairplan Typesetting Ltd., Ware, Herts

Printed and bound in Great Britain by the Bath Press, Avon

Introduction

This book is all about how and why things work. Sometimes it is difficult to learn from words and pictures, so it is also about making things. There are two types of making exercises in the book, **experiments** and **projects**. It is important to do the experiments when they appear, as this will help you to do the projects that follow.

There are also some word games which will help you to remember what has been learnt. When you come across an important word for the first time, it will be printed in **bold** letters. Try to remember these words and what they mean.

In the middle of the book is a game that you can play with your friends. At the back is a Mini-Dictionary which explains the meaning of words that you may not have come across before. If you are not sure about a word, *look it up*.

Introducing Control

What do you think is meant by the word **control**?

What are you doing if you control something?

Why is it important to control things?

Control is a word that you probably hear every day. In order to ride a bike safely, you have to be in control of it. To cook a meal, you need to be able to control the machines in your kitchen.

Control means making something do what you want. In this book you will discover a number of ways of controlling things.

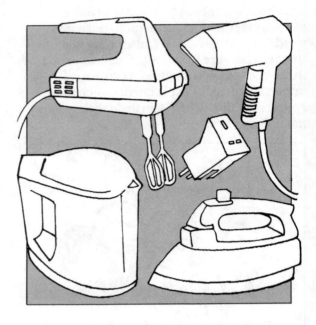

From prehistoric times to the present day, human beings have attempted to control their world. Thousands of years ago, people discovered how to make and control fire, how to make tools for hunting and fishing and how to provide shelter and clothing. As we have developed through the ages, we have learnt how to control both the things we have made and the natural world.

Exercising control over something can be very simple. For instance, how could you control the number of pints of milk left in the morning? Machines often contain many complicated controls but they are designed to be easily used. Can you think of some everyday items that you need to have control over?

Controlling the natural world is becoming more and more difficult. Many of the world's resources are being used up and one day they may disappear completely. As these natural resources are used up, the environment in many parts of the world is changing. Can you think of any examples where people may be losing control over their environment?

Technology and Control

Technology has become a familiar word. Technology is about studying how machines work and using machines to solve problems. We have been doing this for a long time and control is now an important part of technology.

The pictures show some of the things that we all have a certain amount of control over. Can you think of any more?

Air

Water

Electricity

Heat

Technology might also be thought of as the 'appliance of science'. By applying the principles of science found in this book to your designing and making activities, you should be able to develop better and more efficient products. In turn, this will help you to understand how and why science plays such an important role in our lives. Remember that it is impossible to design anything without using technology and that successful technology relies on good design.

Types of Control

Many everyday things can be controlled in only one way. For instance, a **switch** can turn a light off or on. Can you think of some other examples?

Other devices, however, are controlled in a number of different ways: for example, riding a bicycle around an obstacle course.

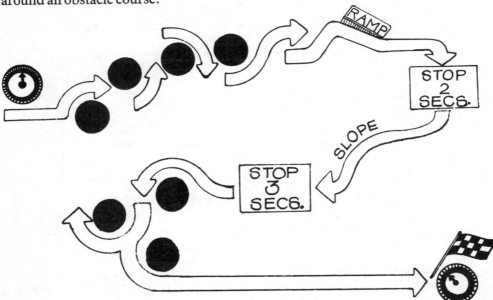

If you have the facilities, you could design a course and try it out. **Get an adult to check your course before you use it. Wear protective or thick clothing.** Make a list of all the control skills that you would need to use in order to tackle the obstacle course. You can check them against the list on the next page.

Have you thought of any more?

The bicycle is a type of **system**. If a number of control devices are used together, they can be described as a system. What other control systems can you think of? Explain why they are systems.

1 Experimenting in Control

Try the following experiments on your own or in a group.

Experiment 1

This is an old idea that uses energy from an elastic band. Make the cotton-reel 'tank' and answer the questions.
You will need:
- some elastic bands
- a cotton reel
- a 50 mm length of 3 or 5 mm dowel
- a matchstick

Questions
1 How is the tank's speed controlled?
2 Can you get it to speed up/slow down?
3 Can you make it turn in a circle?
4 What is the smallest/largest circle the tank will make?
5 Would it help to replace the dowel with a piece of stiff wire to help control direction? Why?
6 How else can an elastic band be used as a control mechanism?

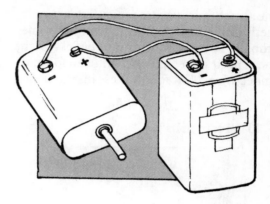

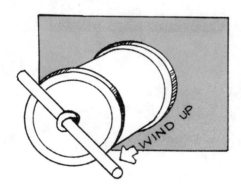

Experiment 2

You will need:
- a motor (1.5–4.5 V)
- a battery
- a battery holder
- a length of wire
- card
- paper fasteners
- paper-clips

Connect the motor to the battery (as in the picture) and answer these questions.

Questions
1 Which way is the motor turning?
2 Can you make it reverse?
3 Can you make a switch that will turn it off and on?

Bicycle Skills List

Observing	Speed
Balancing	Starting
Steering	Stopping

A Membrane Panel Switch

Some switches are very *sensitive*. They turn something on or off at a touch. Think of some examples. To make a membrane panel switch, you will need some thin card, kitchen foil and stranded wire. Now follow these instructions.

1 Fold the card in half.

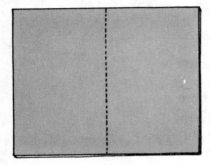

2 Stick two strips of foil on to one half of the card as shown. *Do not cover the foil*.

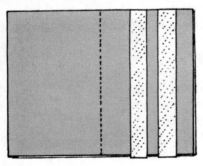

3 Attach the stripped ends of each wire to a foil strip with tape or a staple.

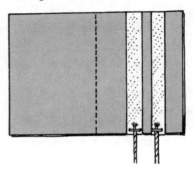

4 Stick a piece of paper, with a window cut out, over the foil strips. These must show through the window.

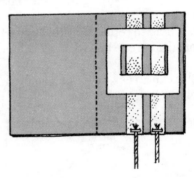

5 Stick the foil **contact** pad on the other side of the card so that it goes over the window when the card is closed.

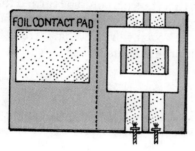

6 Connect the wires to a battery and motor like the ones used in Experiment 2. Close the card and touch the front of it. What happens?

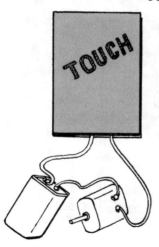

The foil contact pad is pushed up through the window and it connects with the two tracks of foil.

Out of Control

As you have seen in the experiments, control is an important part of our project work here. It is no good switching something on if you cannot easily switch it off again, or making something travel in a straight line when you want it to turn a corner!

Sometimes we lose control of things. Have you ever switched on a light only to hear a faint 'clink' as the bulb blows and you are left in the dark? Or turned on a tap to find that the washer has worn out?

You may have come across things which are difficult to control because they have been designed badly. Can you think of any? It is important that a designer understands how a product will be controlled and by whom.

When doing the projects in this book, think carefully about your ideas and how they will work. Make models and test things before going on to make the final version.

2 Using Wind for Energy and Control

Controlling the Wind

The wind has been used for energy for thousands of years. Sailing-boats were probably the first devices to be powered and controlled with the help of the wind, although early ones relied on oars as well. For many centuries, the sailing-boat provided countries with a form of transport, developing trade and communication links with other countries.

Another invention to use energy from the wind was the windmill, which has been around for over a thousand years. Many mills were built on a pivot so that the sails could be turned to face the wind. Look at the diagram of the windmill. Can you see how it works? What were windmills originally used for?

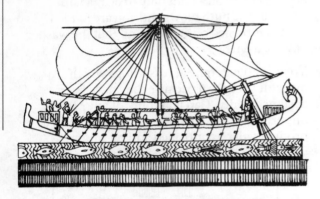

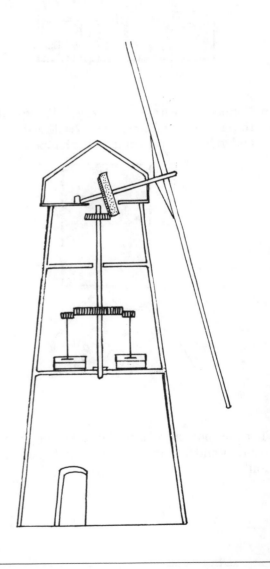

Fact File

The earliest sailing boat was probably a hollowed-out tree trunk which had a leafy branch attached to it. The branch would have been the mast and the leaves would have acted as the sail. This type of sailing boat existed in the Palaeolithic Age. Can you find out when this was?

Experimenting with Sails

A modern development of the windmill is the wind turbine, which is used to generate electricity.

Many popular activities, such as land yachting, kite flying, gliding and sailing, all rely on machines that use the energy of the wind.

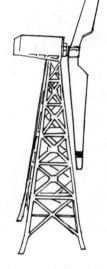

Many machines that use the wind have sails. The wind acts as a force on the surface of the sail and pushes against it, causing movement.

Experiment 3

You will need:
- a 100 mm × 50 mm piece of corriflute (thick card will do if you cannot get this)
- 3 cocktail sticks
- 2 plastic drinking straws
- some sheets of thin paper
- 4 plastic wheels

Make up a simple wheeled chassis, as shown in the picture, and use one of the cocktail sticks as a mast. Trace the sail shapes shown on page 14 on to the paper and cut them out. Make sure that the vehicle moves freely and attach the sail with tape. The sails can be tested one at a time with a hairdryer or a fan.

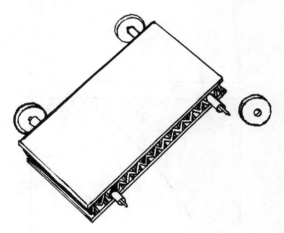

When experimenting, it is important to use a *fair test*. In this case all the vehicles should be exactly the same size, be made from the same material and use the same type of wheels. The sails should be cut from the same type of card and the fan or hairdryer used for the test must operate from exactly the same spot each time.

This will ensure that it is only the different sail shapes that affect the performance of the vehicle.

Testing

Devise a *fair* test to decide:
● Which one moves the fastest?
● What happens if the angle of the sail is changed?

Would the sails work better if they were made from a different material? What materials do you think would be suitable? Why?

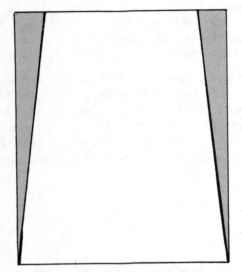

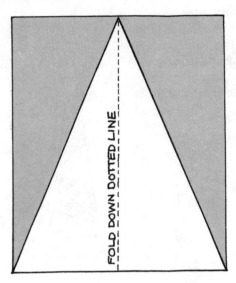

FOLD DOWN DOTTED LINE

Yacht sails tend to be made from a cotton or canvas material that billows out in the wind.

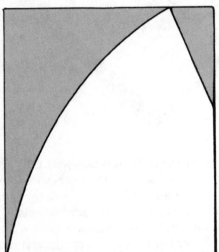

Can you design a better sail? Does the size of the sail affect the speed of the vehicle?

Look at the sail shape in the picture.

The wind pushes against it and it inflates. The air pushing against the sail is forced to slow down and this causes an area of **high pressure** to build up there. The front of the sail billows out and forms a curved surface. Air flowing over this surface moves quickly and creates an area of **low pressure** in front of the sail. This change in pressure helps to push the vessel forwards.

Windsurfers

One of the simplest sailing vessels to use the power of the wind is the windsurfer. It consists of an oval raft which has a small keel attached to it. A curved bar wraps around the triangular sail. By pulling or pushing on this bar, the sail can be moved in any direction to catch the wind. Windsurfing has become a very popular sport, particularly in areas where reservoirs can be used for recreation.

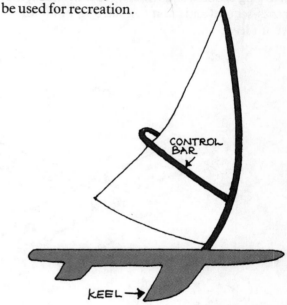

Using Wind for Energy and Control

Land Yachts

Use what you have found out to help with the
following two projects.

Project 1

Design and make a land yacht which can be
used for carrying an egg safely across a distance
of 10 metres. Make up the model shown from
jelutong or softwood which is 12 mm square in
cross-section and use it as a starting-point for
your ideas.

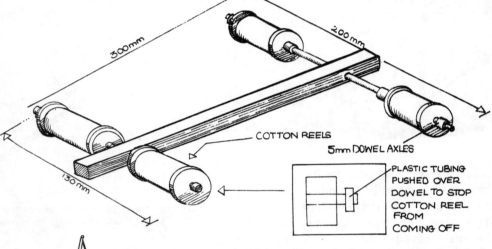

300mm

200 mm

130mm

COTTON REELS

5mm DOWEL AXLES

PLASTIC TUBING
PUSHED OVER
DOWEL TO STOP
COTTON REEL
FROM
COMING OFF

If there is no wind, test your land yacht by using
a hairdryer or a vacuum cleaner which can be
switched to blow.

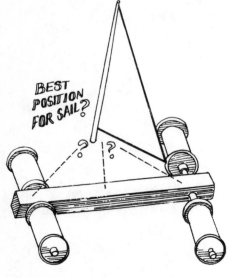

BEST
POSITION
FOR SAIL?

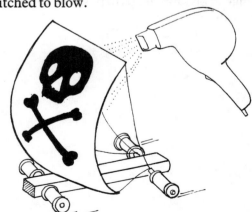

After Testing

Improving the Design

If the wheels do not move freely or the yacht tends to skid, you may have problems with **friction**. Any surface rubbing against another one causes friction. Friction wastes energy, but it can be very useful. If the wheels on your vehicle are not turning freely, check that they are not rubbing against the axles or the frame.

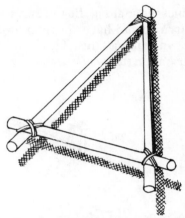

Elastic bands can be used for fixing a frame together quickly for testing.

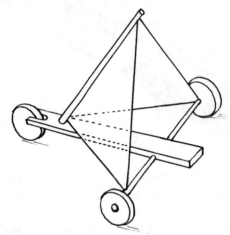

Do you think that the shape of this sail will help the yacht to travel faster? Try it and see.

How important is the choice of material for the sail? Polythene sheeting and bin liners make good sails. Why do you think this is?

Can you design your vehicle so that one sail is easily changed for another one? This would help you to test different shapes and materials quickly.

If the vehicle is skidding, have you thought of adding tyres to improve its grip? Foam lagging for pipes is ideal for fixing on to cotton reels. Is the vehicle slow? If you have added a lot of extra material to your vehicle, it may take more than the force of the wind to move it. Would it help to change the size or shape of the sail? Can you improve the design? Elastic bands can be used for fixing sections of wood together quickly. Use them to experiment with different-shaped frames. How could you steer the yacht? Would three wheels be better than four?

3 Using Air for Control

Floating on Air

Some machines move along, floating on a cushion of air. You may have a hover mower for cutting the lawn that does this. The first machine to use air in this way was the **hovercraft**.

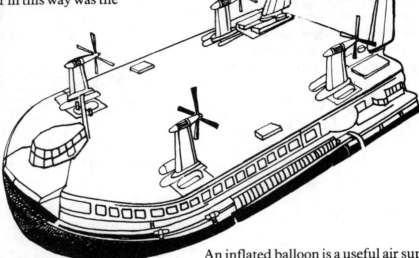

The hovercraft floats along on a cushion of air, allowing it to travel on both land and water.

An inflated balloon is a useful air supply and has the advantage that it can be mounted on board.
Make the model hovercraft shown on the next page and enter the great hover challenge!

In a hovercraft, powerful turbine fans draw in air at the top and force it down underneath. Here a skirt stops the air from escaping in all directions and, as the pressure builds up, the craft rises and floats on a cushion of air.
A hovercraft has large propellers mounted on its deck which drive it forwards. Hovercraft can travel across land and water at high speeds because they are almost free from friction. Why is a hover mower so easy to mow the lawn with? It is possible to use a number of air supplies to drive a model hovercraft. Hairdryers, fans, vacuum cleaners and human lungs have all been used with some success.

A development of the hovercraft is the hover mower, which takes the effort out of mowing the lawn.

Model Hovers

Project 2: Making a Hovercraft

1 Cut out a disc from acrylic or high-density sheet polystyrene.

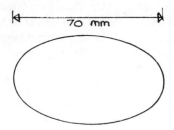

2 Centre drill a 30 mm length of 5 mm aluminium rod, using a 2 mm drill bit.

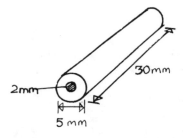

3 Centre drill a piece of 25 mm dowel rod, using a 5 mm drill bit. Cut it to 25 mm in length.

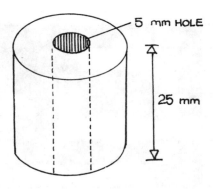

4 Push the rod into the dowel and push the disc on to it, using a file to make the surface level.

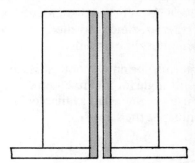

5 Pull a balloon over the dowel and blow through the hole in the disc.

Note: If the rod is pushed in slightly, a piece of 5 mm tubing could be inserted to blow through.

Now that you have the idea, can you adapt your model or design a better one for the great hover challenge?

Using Air for Control

Racing the Hovers

The Great Hover Challenge

Classes:

1 The fastest hover over a distance of 3 metres.
2 The furthest distance travelled.
3 The straightest line travelled.

The course must be on a smooth surface with a 10° downhill angle for the first metre.
You may use any materials available, but all the balloons must be the same size.

You may be able to design a simple board game using the hovers. Acrylic sheet would provide a good surface for this purpose. Why?
Thin card, drinking straws and coloured paper are some of the materials that could be used to style your hover. All these materials are very light in weight. Why is this important?

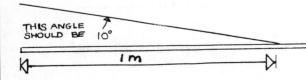

THIS ANGLE SHOULD BE 10°

1 m

Fact File

A development of the hovercraft is the hover bed. It supports the body on a warm cushion of air and is used in hospitals for patients suffering from severe burns.

Can You Remember?

Across

3 Would a land yacht move without it?
6 A type of 8 across.
7 This can waste energy.
8 Fast-flowing air causes this to be 6 across.

Down

1 If you lose it you could be in trouble!
2 Turn something 5 down.
4 To float on air.
5 Not off.

Energy and Work

Using Stored Energy

If you blow into a balloon, it **expands**, and if you pinch the end of it, the air stays trapped inside.

What happens if you let it go?

When it is inflated, the balloon contains potential energy.

When the air is allowed to escape, it rushes out and the balloon shrivels up. The energy released as this happens is called **kinetic energy**. This is the energy of movement. By releasing a balloon in this way, you are wasting energy; but it is possible to use the energy in a controlled way to provide a source of power.

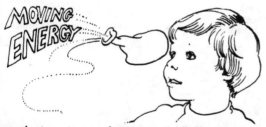

Energy that is causing something to move is called kinetic energy.

Use the model that you made on page 13 for testing the sails to demonstrate this. Replace the mast with a piece of PVC tubing (5 mm bore is ideal) which has a balloon taped to it. Attach this to the chassis and blow up the balloon. What happens when you let go of the model? What would happen if you changed the size of the tubing?

The air squashed in the balloon is a source of energy. Until it is allowed to escape, it is called **potential energy** – that is, energy which is stored. The material of the balloon also contains potential energy, just as a stretched elastic band does, too.

Scientists use the word **work** to describe energy that is being used. An inflated balloon has the ability to do work and so it is said to contain potential energy. Kinetic energy is used when the energy in the balloon is released and it is actually doing work.

M 394

Using Air for Control

Jet Propulsion

Project 3: 'Balloon Fliers'

A competition is being held to design and make a model of a vehicle driven by a balloon. The judges are awarding marks for:

- distance travelled
- speed
- style and construction

The vehicle must not be more than 200 mm long and 100 mm wide. Any available materials may be used.

Give careful thought to:

- **Materials** – light or heavy in weight?
- **Wheels** – large, small, thin or fat tyres?
- **Shape** – streamlining, wind resistance, aerodynamics?
- **Construction** – stability, durability, accuracy?
- **Air outlet** – controlling speed, tubing size?

Look out for the problems shown in the picture below.

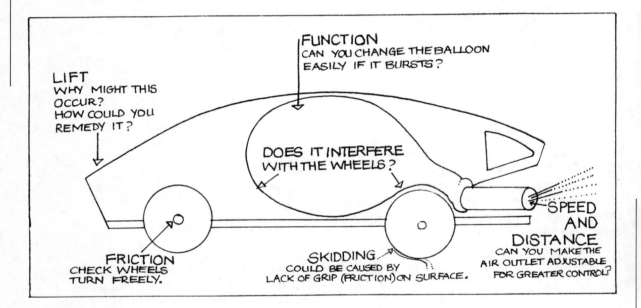

FUNCTION
CAN YOU CHANGE THE BALLOON EASILY IF IT BURSTS?

LIFT
WHY MIGHT THIS OCCUR?
HOW COULD YOU REMEDY IT?

DOES IT INTERFERE WITH THE WHEELS?

SPEED AND DISTANCE
CAN YOU MAKE THE AIR OUTLET ADJUSTABLE FOR GREATER CONTROL?

FRICTION
CHECK WHEELS TURN FREELY.

SKIDDING
COULD BE CAUSED BY LACK OF GRIP (FRICTION) ON SURFACE.

A jet engine works in the same way as your vehicle. An important law of physics states that for every action there is an equal and opposite reaction.

The jet engine releases hot gases which escape under high pressure from nozzles. It is this backwards force, or **thrust** as it is known, which pushes the vehicle forwards.

Fact File

- In 1983 Richard Noble drove at over 633 mph (1012.8 kph) across a desert in North America. His car, called Thrust 2, was powered by a jet engine and broke the land speed record.

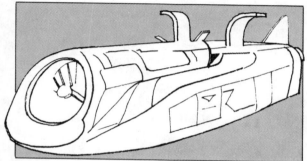

Thrust 2

- A squid uses the same principle as a jet engine to propel itself through the sea. It takes in water through its gills and squirts it out through a narrow funnel at its front. This shoots the squid backwards through the water!

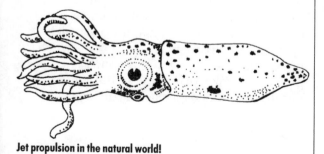

Jet propulsion in the natural world!

Extend Your Knowledge

Acceleration

When you ride a bicycle, you gradually build up speed until you are travelling with very little effort. This gradual build-up of speed is called **acceleration**. In a car or a bicycle, speed can be measured in kilometres per hour from a speedometer. If you can measure the speed of a vehicle, you can find how fast it accelerates. Acceleration is found from the formula

$$\text{Acceleration} = \frac{\text{Change in speed}}{\text{Time taken}}$$

For example:

A bicycle increases its speed from 2 km/h to 8 km/h in 10 seconds. What is its acceleration?

$$\text{Acceleration} = \frac{8-2}{10} = \frac{6}{10}$$

Therefore the acceleration is 0.6 km/h per second.

Using Air for Control

Pneumatic Control

Something in which air is used under pressure is often called **pneumatic**. Common examples are the pneumatic tyres found on road vehicles, the pneumatic drill and all sorts of pumps. When air is **compressed** into a closed space, it behaves rather like a wound-up spring and tries to push back out again. If you sit on a lilo or pump up a bicycle tyre, you can feel this springiness. Bicycles once had wooden wheels with no tyres. Can you imagine what this felt like for the rider!

'LADIES' HOBBY – HORSE BICYCLE

It is easy to demonstrate the springiness of air when it is compressed by making a simple pneumatic system. Take an empty washing-up liquid bottle, a length of PVC flexible tubing (3 mm bore is ideal), a balloon and some tape. Join them together in such a way that the balloon acts as a kind of pump. What happens when you squeeze the bottle?

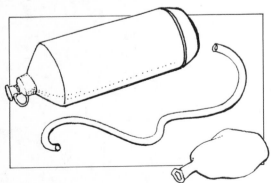

The balloon is made of rubber which allows it to expand as the air pressure builds up inside it. Can you use the pneumatic system to lift something?
Try placing a book on top of the balloon. Does it lift easily? How much does the book weigh? What is the heaviest load that the system can lift?

This type of system is known as a **closed** system. Why do you think this is? The switch shown in the picture works on exactly the same principle. When the bellows are compressed, the air travels through the tubing and is pushed against a rubber skin, or diaphragm, which operates switch contacts. This type of switch is often found in light machinery, such as a photocopier, where it would be difficult to install complicated mechanical linkages.

Many pneumatic systems operate by air fed into them by means of a **compressor**. This provides the system with a constant high-pressure air supply which can be used to open and shut a number of **valves**. This type of system is known as an **open** system and has many uses in industry.

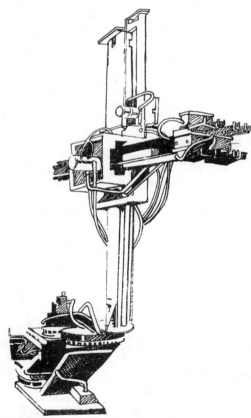

An industrial pneumatically operated machine that uses valves to control the air pressure.

More about Air Pressure

We live in a world surrounded by air. This is called the **atmosphere**. Although you cannot normally feel it, air is constantly pushing against every part of your body and this pushing force is known as **atmospheric pressure**. Luckily, the pressure inside your body is the same as the pressure outside it; otherwise, you would be crushed inwards. When you blow up a balloon, you fill it with air. The air pushes outwards against the rubber skin of the balloon and keeps it inflated. Because the air inside the balloon is compressed, or squashed, the pressure that it exerts is greater than the atmospheric pressure that surrounds it. It is this difference in pressure that enables your simple pneumatic system to support a load.

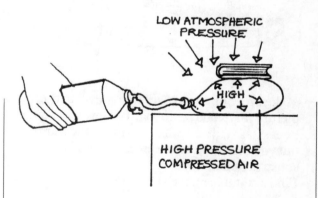

LOW ATMOSPHERIC PRESSURE

HIGH

HIGH PRESSURE COMPRESSED AIR

Pneumatics

Project 4: Pneumatic Control

A toy manufacturer wishes to commission a toy design for an 8-year-old child, which will be known as the 'Snapper'. It is to include a moving feature which should suggest its name. The toy is to be made to the information given, but the moving feature and all customizing is left to the individual designer.

Design and make a prototype that will satisfy the manufacturer's brief and that will make use of the closed pneumatic system.

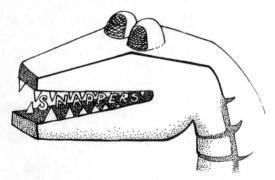

Make up a box construction as shown in the picture. Plywood, chipboard or softwood can be easily glued and pinned to make the frame, but you may wish to use other materials or construction methods.

Find a suitable material for the base.

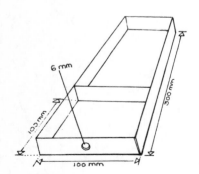

Feed the tubing through the holes in the frame so that the balloon sits into the base.

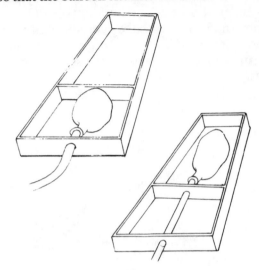

Now you can begin to customize the Snapper!

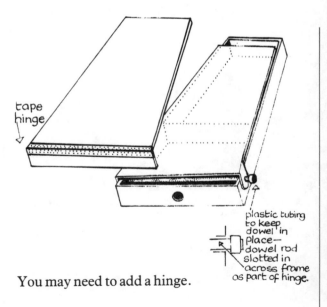

tape hinge

plastic tubing to keep dowel in place— dowel rod slotted in across frame as part of hinge.

You may need to add a hinge.

Building up the Body

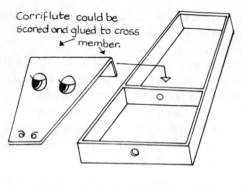

Corriflute could be scored and glued to cross member.

Sides can be made from ply or chipboard.

Corriflute is useful for folding into shape to follow the sides of the structure.

Sides could be cut out of ply or chipboard.

Claws?

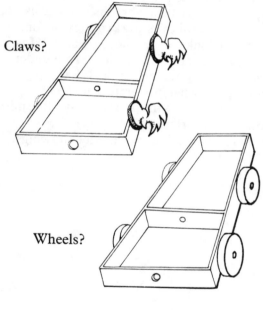

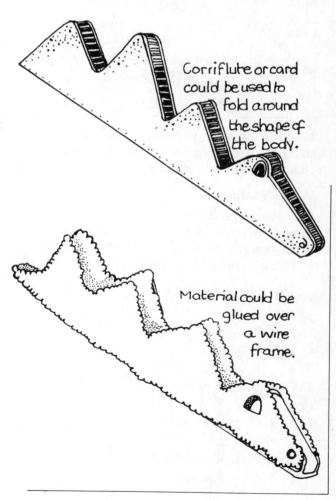

Corriflute or card could be used to fold around the shape of the body.

Wheels?

Material could be glued over a wire frame.

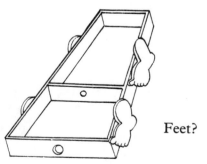

Feet?

Using Air for Control

What is going to move?

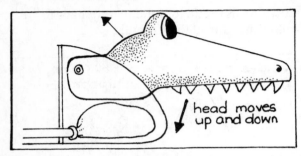

head moves up and down

mouth opens and shuts.

Make sure that the product is finished to a high standard. Remember: to be suitable for a young child it must be safe and durable.

Word Search

In the word search are 14 words that you have come across so far. They may read down, across or diagonally. One word is marked for you, but you have to find it again to make *your* score 14.

```
F T U B I N G A C T
R A C U D S E I L O
I P O T E N T I A L
C E M H N A R A I K
T L P R M U I D R I
I J R U E X W R E N
O H E S T V A L V E
N N S T A Z B R O T
P A S P R I N G P I
S C O N T R O L E C
T I R L E X P A N D
```

Pneumatic Machines

Have you ever heard a deafening noise coming from the direction of road works? It was probably a pneumatic drill in action. This machine uses a supply of compressed air, which is pumped out under high pressure, as a source of power. Unlike the drills in the school workshop, which turn as they bite into a surface, the pneumatic drill is pumped up and down, rather like a hammer. It is sometimes called an air-hammer.

A pneumatic machine that you may find in the graphics studio is an air-brush. This consists of a long narrow tube linking a compressor and a device that looks rather like a pen barrel. Ink is poured into the pen and is sprayed out under pressure through a fine nozzle. It is usually controlled by a hand-operated valve. Airbrushing allows the user to apply a fine mist of colour to a surface and is a very effective method of producing presentation drawings.

4 Mechanical Control

Machines and Motion

Almost all machines depend on some form of mechanical control. They may be simple or complex.

A simple machine.

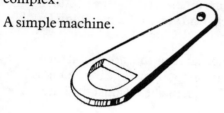

A complex machine.

The bottle-opener is an example of a simple machine. It is a **lever**, which is thought to be the earliest machine. A lever reduces the effort that would normally be needed to move a load. Wheelbarrows, fishing rods and scissors are all examples of levers. A lever always moves around a turning-point, called a **pivot**.

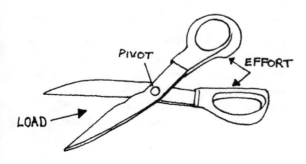

Mechanisms can perform a number of functions. They can change one kind of motion into another kind. There are four kinds of motion:

Linear moves backwards *or* forwards in a straight line.

Oscillating moves backwards and forwards in an **arc**.

Rotary is a circular motion.

Reciprocating moves backwards *and* forwards in a straight line.

Mechanisms can change the **speed**, or **velocity** of motion.

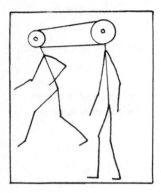

Mechanisms can change the **direction** of motion.

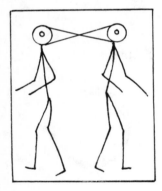

Earlier in this book you thought about the skills needed for controlling a bicycle. This machine has many control mechanisms and gives a useful introduction to learning about how they work.

If you have access to a bicycle, take a closer look at it and examine the parts shown on p.32. This should help you to understand more about the way in which it works.

Cams

Not all mechanisms are found on a bicycle. Here are some others that may be useful in project work. Can you find some examples of them?

Cams are mechanisms that are useful for turning rotary motion into linear motion. As the cam rotates, it pushes against the follower, which rises and falls.

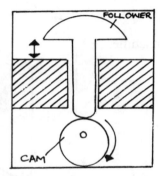

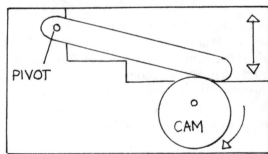

A **plate cam** is a different type of cam. It is rather like a very warped record. As the disc or plate turns, the follower moves up and down under gravitational force. If you watch a needle passing over a record as it plays, you will see it moving gently up and down, in a similar action to the plate cam and follower.

Experimenting with Cams

Experiment 4

Trace the cam shapes on to card or thin plastic sheet and cut them out. Attach them to a piece of card with a paper fastener or eyelet so that they turn freely against it. Cut a follower from a strip of card and use a paper fastener or eyelet to pivot it against the background making sure that when the cam is turned, the follower is pushed against it.

If the angle of the cam when it turns is too steep, it will tend to jam against the follower. When making cam-operated models, it is important to make a mock-up of the cam and follower to check whether or not they work smoothly and produce the kind of movement required.

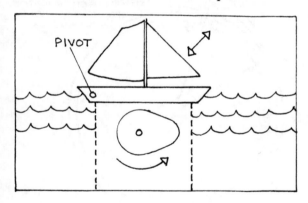

What kinds of movement do the different shapes make?

Can you use the cam to make a moving card?

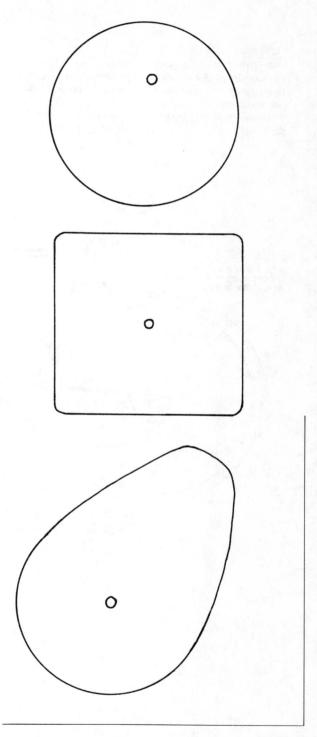

The Bicycle

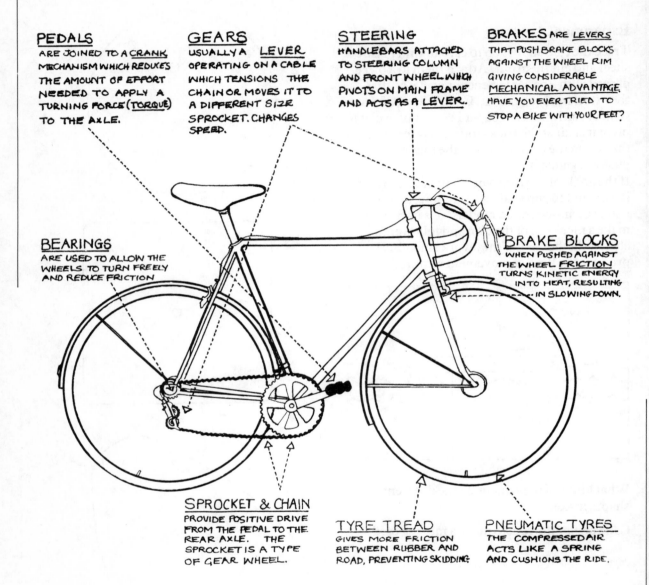

PEDALS ARE JOINED TO A <u>CRANK</u> MECHANISM WHICH REDUCES THE AMOUNT OF EFFORT NEEDED TO APPLY A TURNING FORCE (<u>TORQUE</u>) TO THE AXLE.

GEARS USUALLY A <u>LEVER</u> OPERATING ON A CABLE WHICH TENSIONS THE CHAIN OR MOVES IT TO A DIFFERENT SIZE SPROCKET. CHANGES SPEED.

STEERING HANDLEBARS ATTACHED TO STEERING COLUMN AND FRONT WHEEL WHICH PIVOTS ON MAIN FRAME AND ACTS AS A <u>LEVER</u>.

BRAKES ARE <u>LEVERS</u> THAT PUSH BRAKE BLOCKS AGAINST THE WHEEL RIM GIVING CONSIDERABLE <u>MECHANICAL ADVANTAGE</u>. HAVE YOU EVER TRIED TO STOP A BIKE WITH YOUR FEET?

BEARINGS ARE USED TO ALLOW THE WHEELS TO TURN FREELY AND REDUCE FRICTION

BRAKE BLOCKS WHEN PUSHED AGAINST THE WHEEL <u>FRICTION</u> TURNS KINETIC ENERGY INTO HEAT, RESULTING IN SLOWING DOWN.

SPROCKET & CHAIN PROVIDE POSITIVE DRIVE FROM THE PEDAL TO THE REAR AXLE. THE SPROCKET IS A TYPE OF GEAR WHEEL.

TYRE TREAD GIVES MORE FRICTION BETWEEN RUBBER AND ROAD, PREVENTING SKIDDING

PNEUMATIC TYRES THE COMPRESSED AIR ACTS LIKE A SPRING AND CUSHIONS THE RIDE.

If you have a bicycle, study all its mechanisms and discover how they work. This will help you to understand how many other machine parts operate.

Cranks and Crankshafts

Cranks often look like this.

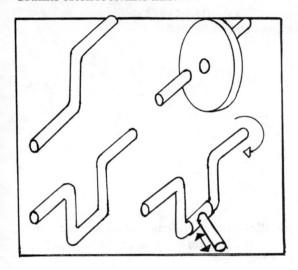

The crank is a special form of shaft or axle.

If a rod is pivoted from the centre of the crankshaft, the rod will move backwards and forwards, as the crankshaft turns.

How are cranks similar to the bicycle pedal? What are they used for? The pistons in an engine connect to a crank.

This mechanism is useful for turning rotary motion into reciprocating motion.

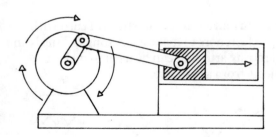

The Sewing Machine

The sewing machine uses both cams and cranks in its system. It is driven by an electric motor, which is connected by a belt to the drive wheel which operates two crankshafts. If you are able to have a close look at one, turn the drive wheel slowly by hand and watch carefully what happens. **Do not put your fingers near the needle**.

A sewing machine not only makes a variety of stitches but is also able to move the material along as it works. A crank moves the needle up and down in reciprocating motion. A series of cams and cranks are attached to the feed-dog, a serrated plate through which the needle rises and falls. The feed-dog moves the material along by rising up and moving forwards between stitches, then dropping and moving back. Different-sized stitches are produced by changing the amount of fabric the feed-dog moves along between each stitch.

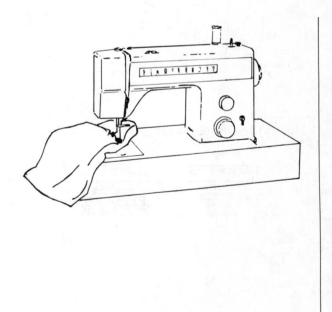

Mechanical Control

Pulleys and Linkages

A **pulley** is a grooved wheel into which a belt can fit. It looks rather like a wheel sandwich!

Pulleys can be used in a similar way to gears, but they need a belt to drive them. Which direction would the driven wheels of the pulleys be turning?

What else do you notice about wheel C? The pulley and belt system is used to transfer motion over a distance and is useful to drive things from an electric motor. It can be used as a simple gear-changing device, such as you may find in the pillar drill.

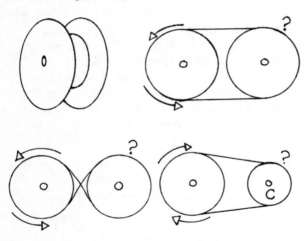

How do you think it works?

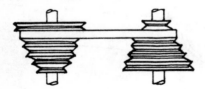

A **linkage** is a number of levers pivoted together. A simple linkage shows how an **input** motion can be reversed:

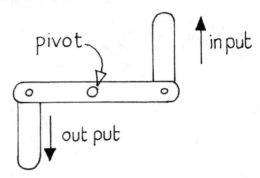

A **parallel motion** is a kind of linkage and can easily be made from strips of card and paper fasteners.
What happens if you push down on points A and B?

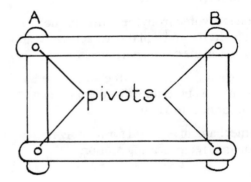

Can you join a number of strips of card with paper fasteners to make a lazytongs? This is an extending arm that can be used to pick up objects from a distance.

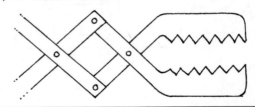

Gears

Gears are wheels with teeth, a bit like the sprocket wheel on a bicycle. When they interlock they give **positive drive**.

If wheels A and B are in contact, in which direction does wheel B turn if wheel A is going clockwise? How can you make the **output** direction the same as the input direction? Gears can be used to speed up or slow down motion. This is known as changing the speed, or velocity. It is done by using gears of differing diameter. The velocity ratio in a simple gear train can be found by counting the number of teeth on each gear.

$$\text{Velocity ratio} = \frac{\text{No. of teeth on driven gear}}{\text{No. of teeth on driver gear}}$$

hand drill use this type of gear. In the diagram, there are 10 teeth on the driven wheel A and 30 on the driver wheel B.

$\frac{10}{30} = \frac{1}{3}$, so the velocity ratio is $\frac{1}{3}$

This is also the gear ratio and is written as 1:3. Gear B turns three times for each turn of gear A.

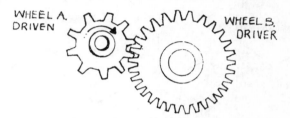

A **worm** gear has only one continuous tooth, like the thread on a wood screw. It meshes with the **wormwheel**, which has its teeth cut at an angle. Because the worm only has one tooth, it is useful when a considerable reduction in velocity is needed: for example, gearing down an electric motor.

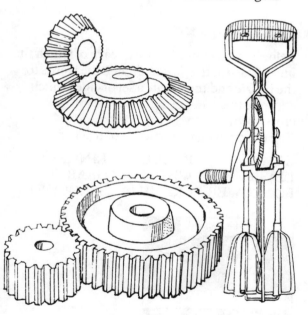

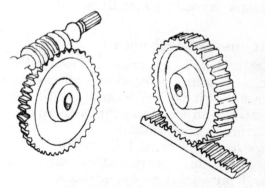

Lego Technic kits have a variety of gear wheels in them. These are very useful in helping you to understand how gears work.

There are many types of gears. **Bevel** gears have their teeth cut at a 45° angle and are used to transmit motion through 90°. A whisk and a

The **rack and pinion** is a type of gearing system. The rack has teeth that are set in a straight line. The pinion is a gear wheel which meshes with the rack as it moves along. It is used to convert rotary into linear motion. Can you find a rack and pinion mechanism on a machine in your workshop?

Mechanical Control

Hidden Mechanisms

Black Box Puzzles

Have you ever looked at something and wondered why it moves in a certain way? See if you can solve these black box puzzles. Copy the boxes and try to draw in the mechanisms that you might find in them.

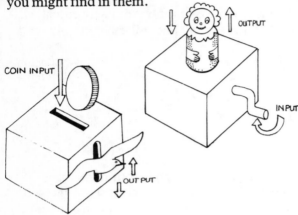

Try to catch out your friends. Design your own black boxes and see if they can work out the mechanisms inside. If you have some Lego Technic, try making a black box.

Motors and Mechanisms

If you did not want to drive a mechanism by hand, an electric motor could be attached to it. Small electric motors that operate on between 3 and 6 volts are ideal for making a variety of electro-mechanical models.
Electric motors are machines that turn electrical energy into mechanical energy. The motor provides a turning force and usually rotates at a high speed. Motors need to be geared down to control the speed of the output. You have seen how to find out the velocity ratio between gear wheels with differing numbers of teeth. If a pulley were to be attached to the motor shaft in order to drive a second pulley linked via a rubber band, how could the velocity ratio be found, given the fact that there are no teeth to count?

The method is similar to that for gear wheels but, instead of the number of teeth, the diameter of the wheel is used. Therefore,

$$\text{Velocity ratio} = \frac{\text{Diameter of driven wheel}}{\text{Diameter of driver wheel}}$$

For example, pulley A has a diameter of 30 mm and pulley B a diameter of 60 mm. Pulley A is the driver pulley. To find the velocity ratio:

$$\text{Velocity ratio} = \frac{\text{Diameter of driven wheel}}{\text{Diameter of driver wheel}} = \frac{60}{30} =$$

Written as a ratio, it is 2:1.

Which Word?

Below is a list of mechanisms, each of which fits into a space in the puzzle. Match the words to the puzzle and then write a set of clues which describe them and turn the puzzle into a crossword.
You can try it on your friends or teacher!

CAM	PULLEY	LINKAGE
LEVER	WHEEL	GEAR
FOLLOWER	CRANK	SPROCKET

A Motorized Machine

Project 5: Pipeline Mole

A mining company needs to lay a cable inside a pipeline. The work will have to be done by a small motorized vehicle which is able to pull the cable through the pipeline. The company have been working on a prototype and they now wish to employ someone to develop the idea further.

You are asked to make up the prototype Mole Mark 1 and develop it through to an end-product that will meet the company's requirements. The vehicle should not be more than 120 mm long and should run from a 3V battery.

Your prototype should be able to pull a length of string a distance of 1 metre through a 100 mm diameter pipe. Use two lengths of guttering, joined with elastic bands, to test your design.

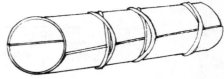

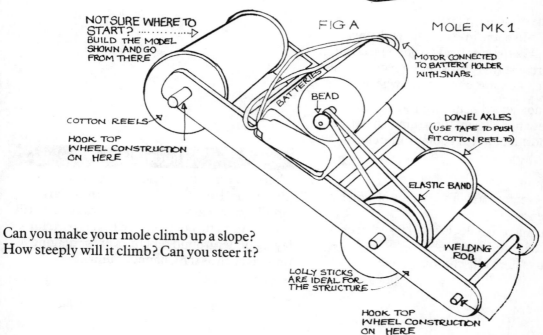

NOT SURE WHERE TO START? ·············▷
BUILD THE MODEL SHOWN AND GO FROM THERE

FIG A   MOLE MK 1

MOTOR CONNECTED TO BATTERY HOLDER WITH SNAPS.

BATTERIES

BEAD

COTTON REELS

HOOK TOP WHEEL CONSTRUCTION ON HERE

DOWEL AXLES (USE TAPE TO PUSH FIT COTTON REEL TO)

ELASTIC BAND

WELDING ROD

LOLLY STICKS ARE IDEAL FOR THE STRUCTURE

HOOK TOP WHEEL CONSTRUCTION ON HERE

Can you make your mole climb up a slope? How steeply will it climb? Can you steer it?

Mole Mark 1

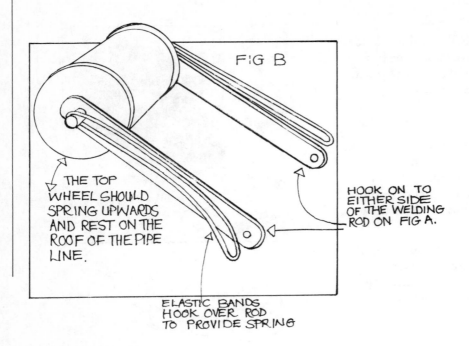

FIG B

THE TOP WHEEL SHOULD SPRING UPWARDS AND REST ON THE ROOF OF THE PIPE LINE.

HOOK ON TO EITHER SIDE OF THE WELDING ROD ON FIG A.

ELASTIC BANDS HOOK OVER ROD TO PROVIDE SPRING

Because the drainage pipe has a curved surface, the wheels will come into contact with only a small area. Foam tyres made from pipe lagging might overcome this problem, but you may like to try and make some customized wheels.

If you cannot get a pipeline for the mole to travel through, make up the vehicle shown on page 37 and adapt it to try some different challenges.

For example:
1 Can you get it to travel a distance of 3 metres in a given time?
2 Can you make a mole that will travel over grass or a bumpy surface?
3 Can you make a hill-climbing mole?

Fact File

There is a tunnelling machine called a 'mechanical mole'. The machine has a large circular cutter at the front that turns like a drill and burrows through the earth. It is used for laying underground pipelines or for digging tunnels such as the Channel Tunnel.

The Bicycle Game

A Game to Test Your Memory

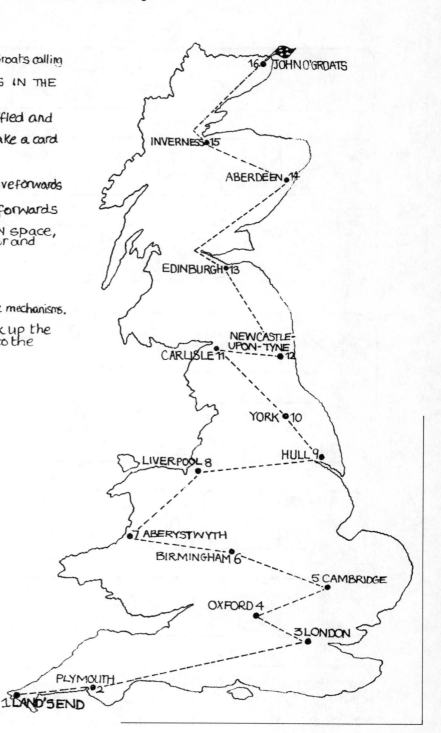

OBJECT
To get from Land's End to John O'Groats calling at ALL towns and cities.
YOU MUST PASS THROUGH THE TOWNS IN THE CORRECT ORDER 1-16.

CONTROL CARDS should be shuffled and put face down on the board.
If you land on a control space take a card and follow the instructions.

RULES
1. To start throw a dice and move forwards the number of places shown.
2. From now on you may move forwards or backwards.
3. If you land on a CONNECTION space, say what the letters stand for and explain the connection.
   EXAMPLE

   G = GEAR
   M = MECHANISM

The connection is that gears are mechanisms.
4. If you are right stay put.
5. If you are wrong you may look up the answer but must GO BACK to the nearest town.

16 JOHN O'GROATS

INVERNESS 15

ABERDEEN 14

EDINBURGH 13

NEWCASTLE-UPON-TYNE 12

CARLISLE 11

YORK 10

HULL 9

LIVERPOOL 8

7 ABERYSTWYTH

BIRMINGHAM 6

5 CAMBRIDGE

OXFORD 4

3 LONDON

PLYMOUTH 2

1 LAND'S END

The Bicycle Game

The Board

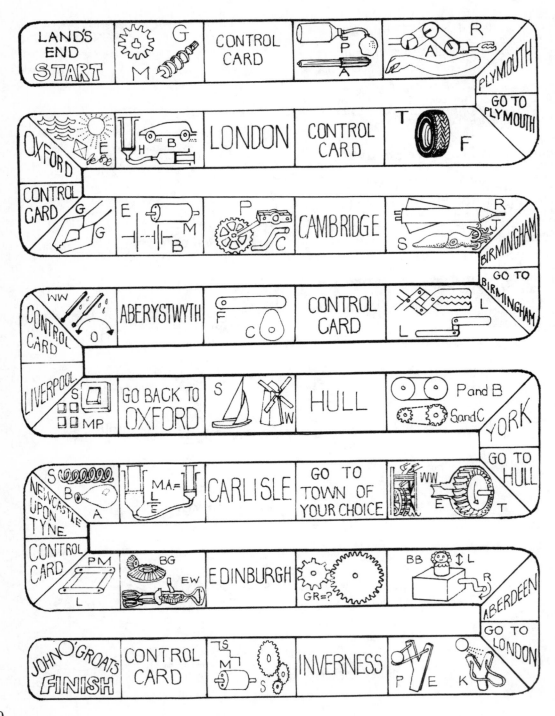

Control Cards

PHOTOCOPY THIS PAGE AND CUT OUT CARDS

CONTROL CARD	CONTROL CARD	CONTROL CARD	CONTROL CARD	CONTROL CARD	CONTROL CARD
NEW TYRES	**10 SPEED GEAR BOX**	**WIN LOCAL TRIALS COMPETITION**	**BRAKES FAIL**	**BUMPY TRACK**	**CHAIN SLIPS**
MOVE TO NEXT TOWN	MOVE TO NEXT TOWN	MOVE TO NEXT TOWN	LAND IN CHANNEL THROW 1 or 6 TO GET OUT	BACK TO LAST TOWN	BACK TO LAST TOWN
IN CONTROL	IN CONTROL	IN CONTROL	OUT OF CONTROL	OUT OF CONTROL	OUT OF CONTROL

CONTROL CARD	CONTROL CARD	CONTROL CARD	CONTROL CARD	CONTROL CARD	CONTROL CARD
PASS ROAD SAFETY TEST	**RIDE SAFELY THROUGH CITY TRAFFIC**	**EMERGENCY STOP**	**WIND AGAINST YOU**	**FORGET PUNCTURE KIT**	**RUN OVER NAIL**
MOVE TO NEXT TOWN	MOVE TO NEXT TOWN	MOVE TO NEXT TOWN	BACK TO LAST TOWN	BACK TO LAST TOWN	BACK TO LAST TOWN
IN CONTROL	IN CONTROL	IN CONTROL	OUT OF CONTROL	OUT OF CONTROL	OUT OF CONTROL

CONTROL CARD	CONTROL CARD	CONTROL CARD	CONTROL CARD	CONTROL CARD	CONTROL CARD
AVOID HITTING BAD MOTORIST	**WIND BEHIND YOU**	**DOWNHILL FOR A MILE**	**PUNCTURE**	**STEEP UPHILL CLIMB**	**SWERVE TO AVOID CAT**
MOVE TO NEXT TOWN	MOVE TO NEXT TOWN	MOVE TO NEXT TOWN	MISS A GO	BACK TO LAST TOWN	BACK TO LAST TOWN
IN CONTROL	IN CONTROL	IN CONTROL	IN CONTROL	OUT OF CONTROL	OUT OF CONTROL

5 Using Water for Energy and Control

Water at Work

Moving water, like air, has been used for thousands of years as a source of energy. Water-wheels are many years older than windmills and were first used for raising water from a stream. The flow of the stream was used to turn the wheel, which had wooden containers joined to it. As the containers turned upside-down, the water spilled into a trough.

A modern version of the water-wheel is the water turbine, which can be used to generate electricity. Fast-flowing water causes the turbine to spin at high speed. The turbine drives large generators which produce electricity. This is the principle of hydroelectric power.

As more became known about mechanisms, the water-wheels improved and were used to turn millstones, pump bellows and trip hammers. Can you think of anything that uses a water- or paddle-wheel?

Water was used as a source of power in many early machines. The water clock was probably the earliest mechanical device to be used for telling the time. It dates back to before the ancient Egyptians but was developed into a more accurate timepiece during the second century B.C.

Another much later development of the water-wheel was the water motor, which became very popular in the late nineteenth century. The motor needed a constant supply of water that was provided by the mains to which it was connected. The water was forced out through a series of small nozzles onto an enclosed turbine. A pulley wheel was attached to the shaft of the turbine and this could be used to drive a variety of both industrial and household machines. Fans, sewing machines and washing machines were just a few of the devices that it could operate. Water motors were used until the end of the First World War when new water rates made them too expensive to run and electric engines and motors began to take over.

Fact File
Hidden in a Welsh valley are the largest man-made tunnels in Europe. They are used to move water to a hydroelectric power station.

Hydraulic Control

In the chapter on air, you found out that air can be squashed, or compressed. Is it possible to squash water in the same way?

Experiment 5

Fill a syringe with water and block the end with your finger. Try to push the plunger in. What happens? Does the plunger move? What happens when you take away your finger?

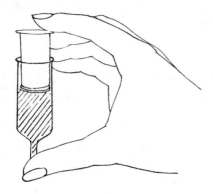

Water cannot be compressed. It will push the plunger upwards as hard as you can push down on it.

Experiment 6

Add a piece of PVC tubing and another syringe to your first one. Join them together as shown. Fill the system with water and try not to let any air in. Push on one syringe. What happens to the other one? Can you explain this?

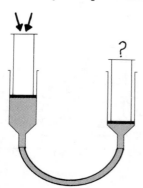

Helpful Hint

To stop air getting into the system:
1 Fill syringe A with water.

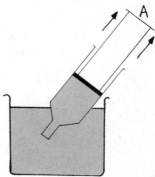

2 Put the tube on to syringe B and fill it with water.

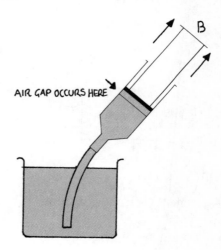

AIR GAP OCCURS HERE

An air gap has been created in syringe B, and it is important to get rid of this.

Hydraulic Control

3 Push the free end of the tube on to syringe A.

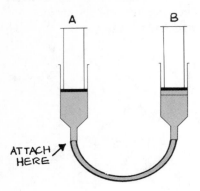

4 Take off syringe B and empty it, making sure that the plunger is pushed right in.

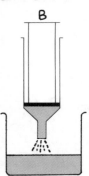

5 Put syringe B back on to the tube.

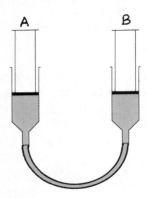

The name given to a system that uses water or another liquid in this way is **hydraulic**. Many machines use hydraulic controls, but they are filled with oil rather than water.

Car brakes use hydraulic controls —.

to stop in a hurry!

Hydraulic controls can be used for lifting . . .

. . . as in a tipper truck.

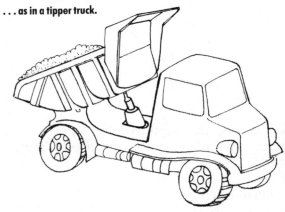

or grabbing!

. . . as in a mechanical shovel.

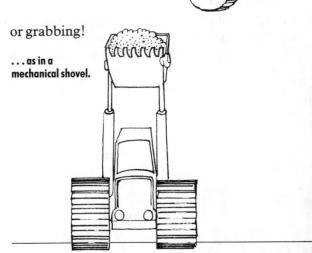

Mechanical Advantage

Some machines give something called **mechanical advantage**.
A simple hydraulic system can help you to understand this.

Experiment 7

You will need a 5 ml and 10 ml syringe for this experiment.
Look at the diagram, then answer these questions:

1 Which plunger would be easier to push down on – the 5 ml or the 10 ml one?
Make up the system and try it out.
Were you right?

2 If you wanted the system to lift a weight, which plunger would you
(a) put the weight on?
(b) push to move it with the smallest amount of effort?

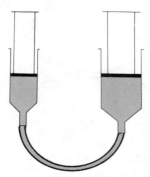

Think back to the previous question before making up your mind. If you can find a picture of a mechanical digger that uses a hydraulic system, this might give you a clue.

Try this experiment by supporting each syringe in a clamp or vice in turn. A sticky pad put on top of the plunger will make the weight more secure, but **watch your toes!**

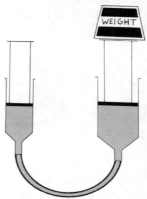

If you want to be more accurate in this test, get hold of a newton-meter. If there is not one available, ask your science teacher. The newton-meter is a kind of spring balance that measures force. Force is measured in **newtons**. Where do you think the term comes from?

Design a test rig that will enable you to read exactly how much force needs to be applied to each syringe to lift the weight.

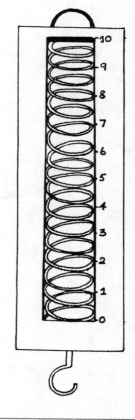

The newton-meter

Were You Right?

It is easier to lift the weight by supporting it on the 10 ml syringe and pushing on the 5 ml one. Did you predict this?

If the force pushing down on the 5 ml syringe was a 1 kg weight, the 10 ml plunger would be able to lift 2 kg.

Assume that the surface area of the 10 ml syringe is twice that of the 5 ml one. The pressure of the water in both syringes is the same, but the 10 ml syringe plunger has twice as much surface for the water to push against. This means that a load on the small plunger can produce a much greater lifting force on the large plunger.
This is an example of mechanical advantage.

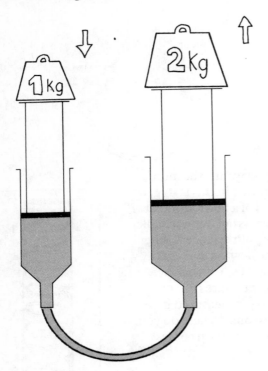

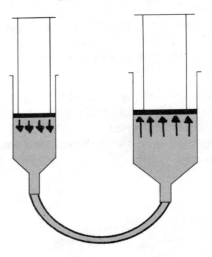

What do you notice about the distance moved by the two plungers? Can you name a mechanism that works in a similar way?

More Calculations

Mechanical advantage can be found by using the formula

Mechanical advantage = $\dfrac{\text{Load}}{\text{Effort}}$

In the experiment, the load = 2 kg and the effort = 1 kg.

$\dfrac{\text{Load} = 2\,\text{kg} = 2}{\text{Effort} = 1\,\text{kg} = 1}$

So, in this system, the mechanical advantage is $\dfrac{2}{1}$ or 2:1.

The syringe that provides the effort is called the **master**.
The syringe that moves the load is called the **slave**.
In hydraulic controls, the syringe case represents the **cylinder** and the plunger, the **piston**.

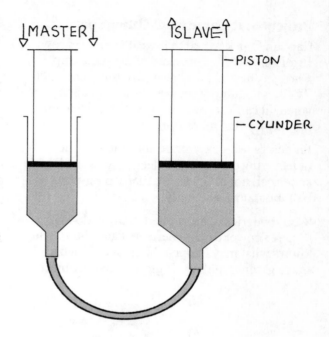

Experiment 8

You have been trying to keep air out of the system. If some has got in by accident, have you noticed a difference in the system? Let some air in and see if you can feel the difference. Does it feel spongy or springy? Can you explain why?
Remember: Air can be squashed and it behaves rather like a spring. Why would air be dangerous if it got into the hydraulic controls of a car's braking system?

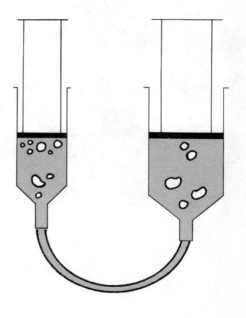

Using Water for Energy and Control

Hydraulics in Action

Project 6: It Came from Outer Space . . .

Captain Quirk wiped the sweat from his brow. In all his years in command of the spaceship *Energize*, he had never seen anything like it. 'Well, one thing's for sure,' he said in a shaky voice, 'it can't stay there. Any bright ideas on moving it, Mr Smock?'

Smock gave a sickly smile and stared at the object lying on the flight-deck. It was small and spherical, not much larger than a ping-pong ball, but far more deadly.

'Could be tricky,' he replied. 'Any sudden surge of power might energize it and then – who knows what may happen?' His face turned a shade greener in the pale glow from the object.

'The hydraulic grab,' said a voice behind them. 'It can be operated safely from a distance and requires no external power supply.'
'Worth a try,' nodded Smock in agreement, his ears twitching at the thought.
'Why do women always have the best ideas?' muttered Quirk under his breath . . .

Can you design and make a hydraulic grab that will move the 'thing' from outer space safely from one place to another? You must never get closer than 300 mm to it.

A Hydraulic Grab

If the 'thing' is the same size and shape as a ping-pong ball, you could design your grab with this in mind.

Here are some ideas to get you started. You will need some card and some paper fasteners for experimenting with.

Any available materials can be used for the solution.

Trace the shapes on to card, cut them out and fasten them as shown.

Push or pull at the points, where the arrows are and see what happens.

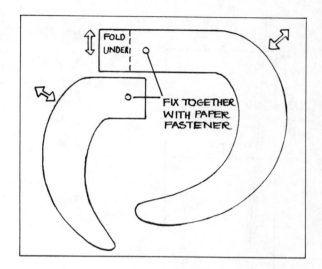

FOLD UNDER

FIX TOGETHER WITH PAPER FASTENER

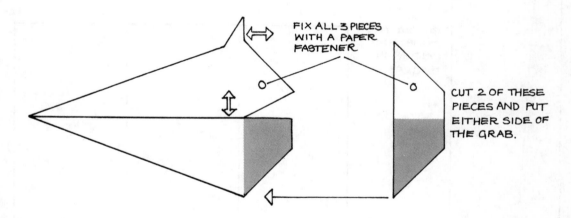

FIX ALL 3 PIECES WITH A PAPER FASTENER

CUT 2 OF THESE PIECES AND PUT EITHER SIDE OF THE GRAB.

Do the arrows give you any clues as to where the plunger of the syringe could be fixed?

Note: Elastic bands and masking tape are useful for fixing the syringe for testing.
A hot-melt glue gun is ideal for permanent fixing.

A Hydraulic Grab

None of these models would pick up a ping-pong ball, so if you use the ideas you will have to adapt the design. Use some card to make up your own ideas, taking into account the shape and size of the ball.

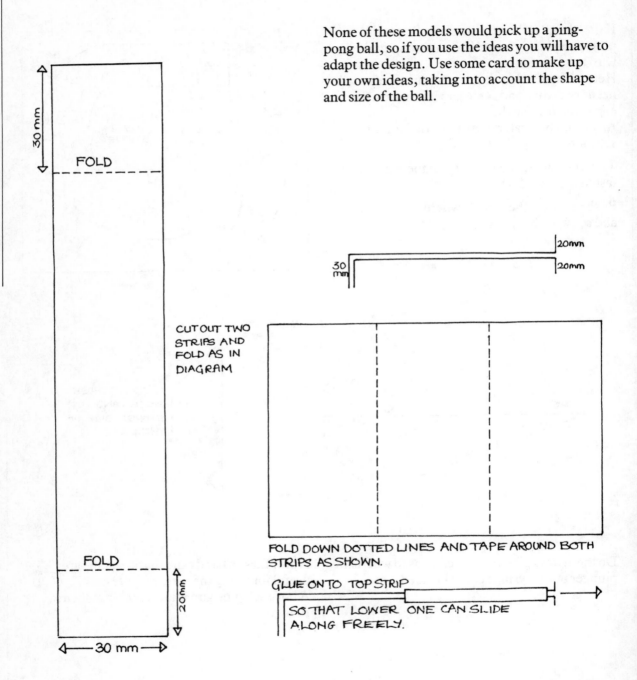

30 mm

FOLD

CUT OUT TWO
STRIPS AND
FOLD AS IN
DIAGRAM

FOLD

20 mm

30 mm

20mm

20mm

30 mm

FOLD DOWN DOTTED LINES AND TAPE AROUND BOTH
STRIPS AS SHOWN.

GLUE ONTO TOP STRIP

SO THAT LOWER ONE CAN SLIDE
ALONG FREELY.

A Hydraulic Grab

Think about the materials you could use for making the grab.
The claw could be made from sheet material. What kind of sheet material can you find? Which type would be the most suitable to use? Why?

Why would acrylic sheet be suitable for making the slide?
What process could you use to form the shape?

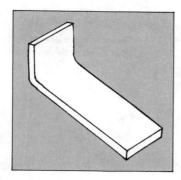

If the grab were made from two hollow jaws that closed around the object, how could they be formed?
Whatever materials and processes you use for the grab, think about how an arm can be attached to it and what this might be made from. How will you hold it? Will it be easy to control?

Hydraulically controlled arms, such as this, are often used by disabled people who find it difficult to pick things up from the floor or reach up to shelves.

Word Search

Hidden in the word search are some words that you came across in the section on water. One has been found to start you off.

```
B L A D H C P I S T O N A R O B T E S
R O R C Y L I N D E R O I Z A M O P Y
T A D I D O U Q U R T I R O N A L V S
S D Y N R I F O R C E L A B E S W R T
M E C H A N I C A L A D V A N T A G E
A S P L U N G E R A L S L A V E T E M
N O P E L A D P E F F O R T Q R E A R
T U R B I N E D F E G I U C R O R L E
E R A S C R A S U R F A C E A R E A D
```

Can you find 16 more?

51

6 Robotic Control

Robot Arms

Robots are often designed to replace people doing dangerous or repetitive jobs. Some robots look like people . . .

others look like animals . . .

but all of them rely in the end on humans for control.

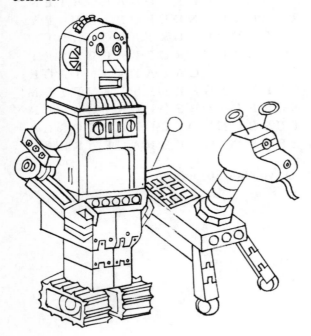

Robot arms used in industry are sometimes called **manipulators**. They can be used for jobs such as paint spraying, sanding, polishing, spot-welding and checking goods. One kind of robot arm in everyday use is a car wash. If you visit one, watch carefully what happens once the programme has started but *don't open the window!*

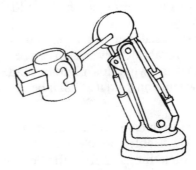

Robot arms can be used for dangerous jobs such as welding . . .

The robot arm is a machine used in many areas. It is similar to your own arm in appearance but has a few advantages. It doesn't get tired,

or need a tea break. It doesn't get bored, or ask for a pay rise. It can work faster and more accurately than a person. It can do dangerous and difficult jobs.

Robot Arms

There may be a robot arm at school which is a scale model of larger industrial manipulators.

Compare your arm to the robot arm.

They look alike and behave in much the same way. Make these movements with your own arm and see how they compare.

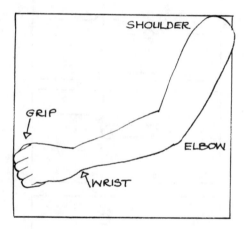

A simple robot arm can make the movements lift, turn and grab.

Lift enables the arm to move either up or down, usually from the elbow.

Turn rotates the arm in a circular movement around the shoulder.

Grab is the movement made by the gripper in order for it to open and close around an object.

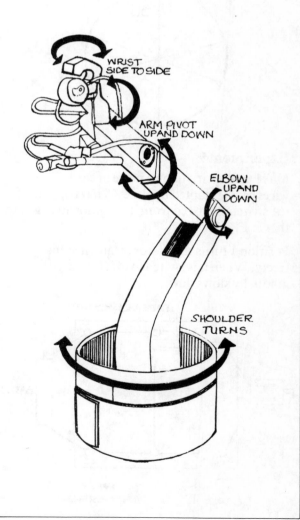

Robot Arms

Many small robot arms use **stepper motors** in their control systems. A stepper motor turns in small 'steps'. If there are 10 steps for every turn of the shaft, each step is 36° of a circle. Because these motors can be made to move in small amounts, they are ideal for driving the robot arm a set distance. To give some idea of how a stepper motor behaves, try this experiment.

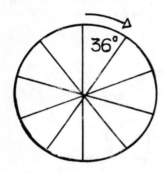

Experiment 9

Make up the circuit on a piece of card. The wires can be stapled in place. What happens if the switch is moved from its position in Fig. 1 to that in Fig. 2?

Position 1 turns the motor on and it spins freely. When it is switched off (Fig. 2), it gradually slows down.

What happens if the switch is moved from its position in Fig. 1 to that in Fig. 3?

If the switch is moved from its position in Fig. 1 to its position in Fig. 3 it stops immediately. By flicking quickly between 1 and 3, you can make the motor move in short steps.

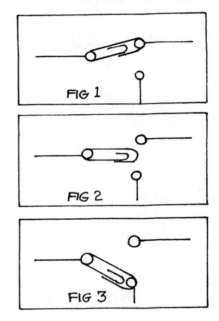

FIG 1

FIG 2

FIG 3

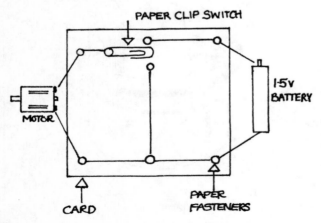

PAPER CLIP SWITCH

1·5v BATTERY

MOTOR

CARD

PAPER FASTENERS

A stepper motor does not work in this way, but the experiment gives an idea of the sort of movement that it makes.

If you attached an arm to the shaft, the movement would be exaggerated. Would it work better if the shaft were not turning as fast? How could you slow it down? Perhaps you could use the circuit in a game of skill or chance. Could it be used as a dice?

Robot Arms

Some industrial robot arms tend to use **servo-motors** rather than stepper motors. They can give a greater turning-force, or **torque**. Servo-motors usually have a built-in gearbox that makes the shaft turn very slowly and increases the turning-force.

Although both stepper and servo-motors are widely used in robotics, many robot arms are powered by a system that you have already looked at – hydraulics. What advantages do you think this system might have over others?

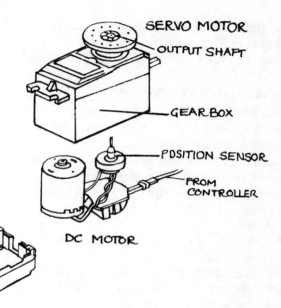

SERVO MOTOR
OUTPUT SHAFT
GEAR BOX
POSITION SENSOR
FROM CONTROLLER
DC MOTOR

Word Maze

Below is a list of words, each of which fits into parts of the maze. The maze is made up of 10 mm square boxes which have been erased. See if you can fit the words into the correct spaces. Use 10 mm squared paper if you need to. With a coloured pencil, turn the maze into a crossword box and write a clue for each word.

ROBOT
LIFT
GRAB
MANIPULATOR
MOTOR
ARM
SERVO
TURN
STEPPER

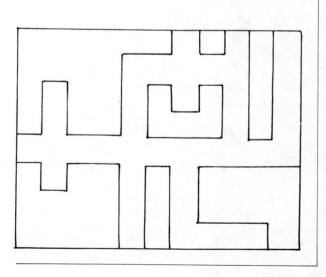

Robot Arms

Note: Before tackling the next project, you must have read the section on hydraulic systems (pp. 43–47).

Project 7: Build a Robot Arm

An educational supplier wishes to market a learning-aid which will introduce primary school children to basic robotics. The product is to take the form of a robot arm which can be easily operated by a child to demonstrate the lift, turn and grab movements. The arm is to be powered by a simple hydraulic system.

You are to:

1 Design and make a prototype which would be suitable for this purpose.
2 'Sell' your idea to the company. This could take the form of producing sales brochures, packaging, research into cost and manufacturing, etc.

You may wish to tackle this project with a partner or a group.

Consider Carefully

- Who will be using the product?
- Choice of materials
- Construction methods
- Safety
- Durability
- Where it will be used

Add any other points to this list that you think are important.

If you are not sure where to start, go back to Project 6 on pages 48–49. If you have made a grab, you are half-way there.
Can you adapt the design of your grab to become the elbow and gripper of the robot arm?

How will the arm lift?

How will it turn?

Look at your solution to the problem. If you were considering buying it, what improvements would you like to see made?

Mini-Dictionary

Bevel gears Gear wheels that are used to transmit motion through 90°.

Cam A mechanism that rotates but does not follow a circular path. A 'follower' mechanism rests on the cam and rises and falls as the cam turns. This mechanism is useful for converting rotary motion into linear motion.

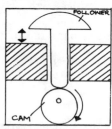

Compressor A device that pumps air under pressure around a system. Compressors are found in most systems that use pneumatic control.

Control Having influence over things; for example, being able to switch something on or off.

Crank A special kind of axle which has been bent or formed through 90° in order to make it easier to turn something. For example, a crank mechanism supports the pedal on a bicycle and reduces the amount of force needed to turn the sprocket wheel.

Crank wheel A wheel with a handle attached to it off-centre, making it easier to turn.

Crankshaft A shaft with more than one crank on it, used for driving a slider mechanism or piston. A crankshaft can be used for converting rotary motion into linear motion.

Cylinder A tube-like container. Cylinders form the outer casing of pistons and have a nozzle at one end.

Energy The ability to do work. Energy cannot be created or destroyed, only converted to another form.

Follower A mechanism that rests on a cam and rises and falls as the cam turns.

Friction This is caused by two surfaces moving against one another. Friction converts moving energy into heat.

Gears Wheels which have teeth and mesh with one another to give positive drive. Gear wheels can be used to change the speed and direction of motion.

Grab A term used to describe the gripping mechanism of a robot arm.

Hovercraft A machine that can travel across land or water on a cushion of air.

Hydraulic Something that uses water or another liquid under pressure.

Instrument wire A thin-gauge wire suitable for use in electronic circuits.

Input What is fed into something, such as energy or information.

Kinetic Energy that is moving.

Lazytongs An extending mechanism made up of linkages.

Lever A mechanism that reduces the amount of effort normally required to move a load.

Lift An up or down movement made by a robot arm. An upwards force caused by an area of high pressure.

Linear Movement in a straight line.

Linkage A mechanism made up from a number of levers that are pivoted together.

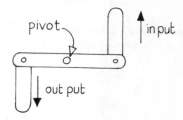

Manipulator A large industrial robot arm.

Master cylinder The piston in a hydraulic or pneumatic system that provides the effort.

Membrane panel switch A sensitive switch that operates from touch.

Motor A device that converts stored energy into moving energy in order to drive something.

Oscillating Movement backwards and forwards in an arc.

Output What comes out of something: energy or direction of motion, for example.

Piston A type of plunger that forces air or liquid through a cylinder.

Pivot A fixed point around which something is able to turn.

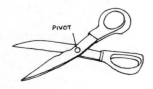

Pneumatic Something in which air is used under pressure.

Potential energy Energy that is stored.

Pressure The force with which something pushes against something else.

Pulley A grooved wheel in which a belt can sit. A pulley system can be used to transmit motion over a distance, and change the speed and change the direction of motion.

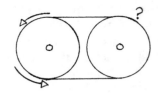

Rack and pinion A gear mechanism that converts rotary into linear movement.

Reciprocating Constant movement backwards and forwards in a straight line.

Robot A machine designed to replace people when they are doing difficult and dangerous jobs.

Rotary A circular motion.

Servo-motor A special type of motor that usually contains a gearbox.

Slave cylinder The piston in a hydraulic or pneumatic system that moves the load.

Sprocket A toothed wheel that interlocks with a chain to provide drive. This mechanism is used on a bicycle to transmit motion between the pedal and the back wheel.

Stepper motor A motor that can be made to turn in small steps.

Stranded wire Several fine strands of wire twisted together and contained within a plastic sleeving.

Switch A device that can be used to control the current flow in an electric circuit. A device that can be used to operate a mechanism.

System A number of control devices linked together.

Technology Using scientific principles and applying knowledge to solve problems.

Thrust A backwards force caused by air or water being released under pressure.

Turbine A fan-like motor that is driven by a flow of water, air or gas.

Turn A movement through 360° (as performed by a robot arm).

Velocity Speed.

Windmill A device that converts energy from the wind to mechanical energy.

Worm gear A gear with only one tooth; similar to a screw thread.

Wormwheel A gear that has its teeth cut at an angle to mesh with a worm gear.

Index

characters created by

lauren child

Snow
is my
FAVOURITE
and my best

PUFFIN

Text
based
on
script
written
by Samantha
Hill

Illustrations
from
the
TV
animation
produced
by
Tiger
Aspect

PUFFIN BOOKS
Published by the Penguin Group: London, New York, Australia,
Canada, India, Ireland, New Zealand and South Africa
Penguin Books Ltd, Registered Offices: 80 Strand, London WC2R 0RL, England

puffinbooks.com

First published 2006
This edition published 2008
1 3 5 7 9 10 8 6 4 2
Text and illustrations copyright © Lauren Child/Tiger Aspect Productions Limited, 2006
The Charlie and Lola logo is a trademark of Lauren Child
All rights reserved
The moral right of the author/illustrator has been asserted
Made and printed in China
ISBN: 978–1–856–13185–8

This edition produced for The Book People Ltd,
Hall Wood Avenue, Haydock, St Helens WA11 9UL

I have this little sister Lola.
     She is small and very funny.
Today Lola is extremely excited
     because the man on the weather
says it's going to snow.

Lola cannot wait for the snow to come.

She says, "Snow is my favourite

and is my best."

I say, "Remember, Lola,

snow can only come when it is very, very cold.

Dad said it might not snow until midnight.

Or even tomorrow."

"I know,"
says Lola,
"but it is extremely
cold right now.
So I think the
snow will come
sooner rather
than midnight."

At bedtime, Lola says,
    "Do you think it has
started **snowing** now, Charlie?"

"No, go to sleep, Lola."

She says, "I can't because
        it might come while
    I'm asleep, **sleeping**.

I'll just do **one more**
check...
    No snow.
    Not **yet**."

"See?" I say.
"Go to sleep."

But a little bit later
I hear Lola creeping
out of bed again.

"Ooooh!" she says.
"It's Snowing!
Charlie, come quick.
It's Snowing, it's really,
really Snowing!"

So I watch the snow with Lola.
She says, "Can we go out
and play in it now?"

"Not now, Lola," I say. "Wait until morning.
Then there'll be more and we can
go on the sledge with Marv and Sizzles.
And you can build a snowman if you want."

In the morning,
Lola shouts,

"Charlie!
Get up, Charlie!
Mum! Dad!

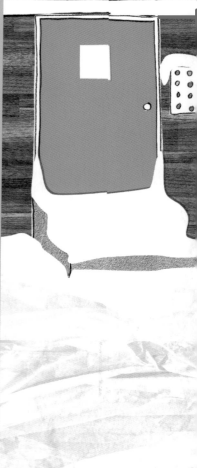
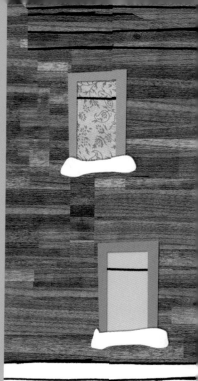

It's all gone
extremely white!"

So Mum and Dad took us to
the park and Lola was right,
everything had turned **extremely**,
completely **white**.

Then we see Marv and Lotta.
   And I say, "Where's Sizzles?"

"Yes," says Lola, "where's Sizzles?"

Marv points to a small pile of snow.
   "He's here!"
      he says. "Look!"

Lotta and Lola
make snow angels.

Lola says,
"Snow
is my
favourite
and my
best."

"I love
snow!"
says Lotta.
"It's my
best
too."

Then we find a big hill and we all
go on the sledge. Even Sizzles!

I say, "Ready?
Steady?
Go!"

Wheeeeeeeee

eeeeee!

Then me and Marv
build a **snowman**.

Lotta says,
"Let's make a
Snow doggy.
Come on, Lola!"

Later we go home to have some hot chocolate.
Marv says, "Mmmm. I love hot chocolate!"
Lola says,
"I love snow. Tomorrow I might put snowdog
and Sizzles on the sledge for a ride."

"I'm going to make a snow kennel," says Lotta,
"... and what about snow puppies?"

"Yes!" says Lola. "We can have lots of snow puppies!"

But when we go to the park the next day,
Lola can't make anything.

"It's gone!" she says.

"All the lovely snow is absolutely gone.
There's no more white, Charlie.
It's all cold
            and wet
   and brown.
            And snowdog's gone."

So we go home again.

Lola says,
"Why can't we
         have **snow**
 every day?"

And I say,
"Because it wouldn't be special.
        Imagine you had a birthday
every day, so you had parties
and cakes and presents
                all the time."

And Lola says,
"What's wrOng with having

                birthdays every day?"

And I say,
"It wouldn't be a treat, would it? I'm not
sure you would like snow every day."

"I would, Charlie," says Lola.
    "Snow is my favourite
    and is my best."

Then I have a really good idea.
"Well, imagine a completely white land...

... where it's snowy and cold every day.
It's called the Arctic."

"Look at the polar bear," says Lola.
"What's he doing, Charlie?"
I say, "He's going for a swim."

"I'd like to go swimming," says Lola.
"Where's the beach?"
I say,
"There isn't a beach, Lola.
It's far too cold for us to go swimming."

Then I say, "And then there's this place right at the very bottom of the world, called the Antarctic, where you get seals and whales and ...."

"Penguins!" says Lola.
"Don't the penguins look smart,
Charlie! They look like they're
going to a party!
I wish I was wearing my best, smartest
party dress, you know, the stripy one."

And I say,
"You couldn't wear your stripy dress in the Antarctic.
You have to wear your coat all the time
because it's so cold."

"Oh yes," says Lola, "I forgot."

Then I say, "But when it's all snowy, you can do this...

And I say, "But why? I thought snow was your favourite and was your best!"

says, "but can we go home now?"

"Come on!"

And I say, "Isn't it amazing?"

And we slide on the ice with the penguins.

"Wow!" says Lola.

"Yes, Charlie," she

Lola says, "I do like it, Charlie. But I'm just a little chilly!"

"Snow is my favourite and my best, Charlie," says Lola, "but if it was snowy all the time there would be lots of things you couldn't do. So we're maybe lucky, we can do swimming and have stripy dresses and have snow.

But I do feel sad that the snow has all gone."

So I say, "I've got a

"A t^eeny weeny sn0wman
who lives in the freezer!"
says Lola. "How did he
get in there?"

"I don't know!" I say.

…rprise for you.
…Don't look round!"

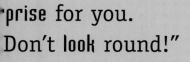

Lola says, "He's me|ting!"
I say, "Shall I put him back
in the freezer so we can keep him?"
"Oh no, Charlie," says Lola.
"Let's watch him me|t!"